CW01508408

1

Identification of Opals

Australian Gemstones Series

Book 9

by Trudy Toohill

Published by Wild Colonial Press
PO Box 7151
Mount Crosby Qld 4306
Australia

jerilderierareandcollectablebooks@live.com.au

First published 2016

OTHER BOOKS BY THE AUTHOR

Non-fiction

Early Australian Gemstone Discoveries

Early Australian Diamond Discoveries

The Physical and Optical Properties of Gemstones

Identification of Quartz and Its Varieties

Identification of Feldspar and Its Varieties

Identification of Amber

Identification of Coral

Identification of Pearls

Australian Gemstones : Australian Gemstones Series Book 1 to 4

The Reporting of Ned Kelly and the Kelly Gang

The Reporting of Ben Hall the Brave Bushranger

The Reporting of Captain Thunderbolt the Gentleman Bushranger

The Reporting of Captain Moonlite the Loyal Bushranger

The Reporting of Australian Bushrangers : Book 1, 2 & 3 of the Australian Bushrangers in Print Series

Australia's Haunted History

Early Queensland Ghosts and Hauntings

Early Australian Ghosts and Hauntings

For

Dad

CONTENTS

INTRODUCTION

This book contains a wealth of information on various types of opals, their appearance, diagnostic properties, imitations and how to tell the difference between potch and precious opal.

Designed for the beginner or experienced, it will greatly assist those wishing to find out more about how to correctly identify opals.

There is also details on the correct readings for opals, from various gemmological instruments, such as spectroscope, polariscope, refractometer, and more, which will aid in accurate identification.

This is book 9 in 'The Australian Gemstones' Series.

I hope you enjoy reading and learning all about opals.

What is the difference between Opal Potch and Precious Opal?

Precious Opal

The spectacular play of colour in precious opal is due to regular stacking of uniformly sized spheres of amorphous silica with regularly arranged spaces between them. The structure of these spheres and spaces form a three-dimensional lattice producing a three-dimensional diffraction grating. The play of colour is due to diffraction of light by the spheres. The colours seen depend on the size of the spheres. Larger spheres give red, while smaller spheres may give only blue flashes of colour, and varying sizes in between.

Opal Potch

In opal potch, spheres of amorphous silica are irregularly stacked, irregularly spaced and irregularly sized. As a result, there is no play of colour and the opal is referred to as common opal, milky opal or potch.

White Opal Appearance and Diagnostic Properties

- Hardness:- 5.5 – 6.6
- Colour:- light or white body colour with a fine play of colour
- Chemical Composition ($SiO_2.nH_2O$)
- Fracture:- conchoidal or uneven
- Polariscope:- single refraction
- Crystal System:- none – amorphous substance
- Refractive Index:- 1.44 to 1.46
- Specific Gravity:- 2.10
- Spectroscope:- absorption spectrum is of little value
- U.V. Light:- under short-wave and long-wave opal varies from white to bluish, brownish or greenish often with a green phosphorescence for approximately 8 – 10 seconds
- Diaphaneity:- transparent, translucent, opaque
- Lustre:- waxy to resinous
- Cleavage:- none

Black Opal Appearance and Diagnostic Properties

- Colour:- black or dark blue, green or grey body colour with vivid play flashes of colour springing from the dark stone
- Specific Gravity:- 2.10
- U.V. Light:- inert
- Diaphaneity:- transparent, translucent, opaque
- Hardness:- 5.5 – 6.6
- Lustre:- waxy to resinous
- Cleavage:- none
- Chemical Composition ($SiO_2.nH_2O$)
- Polariscope:- single refraction
- Crystal System:- none – amorphous substance
- Refractive Index:- 1.44 to 1.46

Boulder Opal Appearance and Diagnostic Properties

- Thin veins of opal ramifying through hard jaspideous brown-coloured boulders
- 2 subvarieties of boulder opal are sandstone boulder and the yowah nut – sandstone boulder consists of concretions made up of shells of coarse sandstone and hard siliceous clay with layers of opal between them or filling the centre – yowah nuts are small boulders about the size of a walnut and are found in a unique formation consisting of a regular band containing the nuts packed in conglomerate, the opal is found either as a central kernel or as thin veils surrounding an ironstone centre or traversing the nut but never reaching to the outer edge

Treated Opal Matrix Appearance and Diagnostic Properties

- Porous cream to grey coloured Andamooka matrix opal has been impregnated by particulate carbon to create black-dyed Andamooka matrix opal. This is manufactured by carbon impregnating of a fine-grained, porous, siliceous South Australian claystone that is included by pinpoints of precious opal.
- Hand lens examination will reveal that this colour enhanced opal consists of pinpoints of precious opal set in a black-stained, fine-grained matrix.
- Specific Gravity:- 2.65 to 3.00

Composite Opal – Doublet – Appearance and Diagnostic Properties

- Thin film of opal exhibiting a fine play of colour and backed with pieces of potch, black onyx, or a black glass called opalite
- Diaphaneity:- transparent, translucent, opaque (with hand lens you can see joins at girdle)
- Hardness:- 5.5 – 6.6
- Lustre:- waxy to resinous
- Cleavage:- none
- Polariscope:- single refraction
- Crystal System:- none – amorphous substance
- Refractive Index:- 1.44 to 1.46

Composite Opal – Triplet – Appearance and Diagnostic Properties

- An ordinary opal doublet is completed with a cover glass of rock crystal which fits over the top of the opal
- Similar triplets of opal made with caps of synthetic colourless spinel and synthetic colourless sapphire can also be found
- Also known as Triplex Opal
- Lustre:- waxy to resinous
- Cleavage:- none
- Diaphaneity:- transparent, translucent, opaque
- Hardness:- 5.5 – 6.6
- Polariscope:- single refraction
- Crystal System:- none – amorphous substance
- Refractive Index:- 1.44 to 1.46

Opal Potch Appearance and Diagnostic Properties

- Specific Gravity:- 1.98 to 2.20
- Diaphaneity:- translucent to opaque (no play of colour)
- Hardness:- 5.5 – 6.6
- Chemical Composition ($SiO_2.nH_2O$)
- Fracture:- conchoidal
- Tenacity:- brittle
- Colour:- occurs in different tints of yellow, brown to black
- Lustre:- waxy to resinous
- Polariscope:- single refraction
- Crystal system:- none – amorphous substance
- Refractive Index:- 1.44 to 1.46
- Cleavage:- none
- U.V. Light:- natural opal phosphoresces for approximately 8 – 10 seconds

Gilson Synthetic Opal Appearance and Diagnostic Properties

- Diaphaneity:- transparent, translucent, opaque
- An early Gilson Synthetic Opal exhibited colour bands rather than patches following the long direction of the oval cut stone (of course this will depend on how the material is cut)
- Various Types:- black, white, water, fire
- A diagnostic lizard-skin effect can be seen within the patches of colour
- U.V. Light:- stones may show a chalky-blue fluorescence with a similar but stronger surface glow with some phosphorescence, however these effects will not be seen in all Gilson opals – the glow is stronger under long-wave ultra-violet light – Gilson Synthetic Opal phosphoresces for approximately 1 – 3 seconds only
- Lustre:- waxy to resinous
- Cleavage:- none
- Chemical Composition:- Gilson opal uses silica spheres, however the composition does vary, some yellowish-brown stones contain a variety of organic compounds and others contain about 0.5% crystalline ZrO_2.
- Gilson Synthetic Hydrophane Opal:- improves in appearance after being placed in water

BIBLIOGRAPHY

Gems Their Sources, Descriptions and Identification by R. Webster

Minerals & Gemstones of the World by G. Brocardo

Minerals of the World by Rudolf Duda & Lubos Rejl

Illustrated Encyclopedia of Minerals by Dr Alan Woolley

Identifying Rocks and Minerals by Basil Booth

Sandstone Identifier – Gems & Precious Stones by Cally Hall

GIA Handbook of Gem Identification, ninth edition by Richard T. Liddicoat Jr.

Dana's Manual of Mineralogy, eighteenth edition by Cornelius S. Hurlbut Jr.

Gem Testing, ninth edition by B.W. Anderson

The Practice Study of Crystals, Minerals and Rocks by Cox, Price and Harte

Elements of Chemistry Volume 1 by R.B. Bucat

Chambers Science and Technology Dictionary by Professor Peter M. B. Walker

A Field Guide to Australian Rocks, Minerals and Gemstones by Wolf Mayer

Physical Principles of Chemistry by R.H. Cole and J.S. Coles

Printed by Amazon Italia Logistica S.r.l.
Torrazza Piemonte (TO), Italy